The Vetiver System for Soil and Water Conservation

Proven and *Green* Environmental Solutions

The Vetiver Network International
USA

2008
First Edition
Published by The Vetiver Network International

The Vetiver Network International is a US tax-exempt non-profit foundation under US IRS code 501(3) (c) dedicated to the world wide promotion of the Vetiver System

Cover by Richard Grimshaw

Preface

This handbook is based on John Greenfield's earlier book - *Vetiver Grass: The Hedge Against Erosion*. It updates the text, and matches actual photographic examples of Vetiver System applications against the original line drawings.

The handbook remains primarily a handbook for "lay" users interested in basic soil and water conservation practices. It is simply written and contains enough information and facts so that even the least technically educated users can confidently read, apply, and achieve good results.

Readers who may want more information about the use of the Vetiver System for soil and water conservation, as well as for other applications, will find a wealth of research data, photographs, power point presentations and links to other technical manuals and publications on the Vetiver Network International's website at www.vetiver.org.

The Vetiver System has come a long way since John Greenfield wrote the first handbook in 1987. However the same basic principles apply, and the technology remains low cost, effective and easy to use. The technology is used in every tropical and subtropical country in the world as well as some countries outside the tropics.

John Greenfield continues to practice what he preaches and is expanding the use of the Vetiver System in his own country - New Zealand.

In 2006 John received the much deserved Norman Hudson Award for his services to agricultural, soil and water conservation, and the vetiver grass technology in particular.

Richard Grimshaw
Chairman - The Vetiver Network International

Forward

In the late 1980s, I wrote a handbook to benefit extension workers in India who were introducing vetiver grass (*Vetiveria zizanioides*, reclassified recently as *Chrysopogon zizanioides*) technology to farmers for the first time. This book is a modification of the extension worker's handbook, first published (1987) in India as "Vetiver Grass: A Method of Vegetative Soil and Moisture Conservation" by the World Bank and later published as "Vetiver Grass: The Hedge against Erosion". The handbook, now in its fifth edition, has been edited and reprinted in many languages, and has become known as "The Little Green Book".

In this new handbook I explain in simple terms what erosion is, how to recognize it, and how to apply the Vetiver System to help prevent it. I also introduce the concept of contour cultivation and moisture conservation, both practices essential to successful sustainable farming in rainfed areas and an essential part of the Vetiver System.

At this time many countries, particularly those in the tropics and semi-tropics, are facing potential food crises. The Vetiver System should be one of a number of technologies to help resolve the problem through conserving soil, nutrients and soil moisture, and farmers should be encouraged to apply the technology on a wide scale.

I would like to thank my colleagues, old and new for their contributions to this handbook, and particularly Dick Grimshaw who edited and published it.

John C. Greenfield
Director
The Vetiver Network International
May 1 2008

Contents

Sheet Erosion	1
Rainfed Farming	4
Vegetative Contour Hedges	13
Following the Contour	16
Establishing Vetiver Hedges	28
Moisture Conservation	34
Why Vetiver Grass Is the Ideal Plant for the Vegetative System of Soil and Moisture Conservation	37
Other Practical Uses for Vetiver Grass	39
Stabilizing terrain and structures	39
Establishing tree crops	41
Mulching	42
Establishing forests	44
Stabilizing masonry walls for hill farming	45
Protecting roads.	48
Stabilizing wasteland development	50
Stabilizing riverbanks and canal walls	52
Protecting dams	55
Common applications	57
Management Tips	59
General observations	59
Selecting planting material	61
Establishing nurseries	61
Field Planting	62
Table 1. Slope, Surface Run, and Vertical Interval	64
Table 2. Cost of Land Treatment with Contour Hedges of Vetiver Grass	65

Editor's note: This handbook contains line drawings and images that are in most cases matched and grouped by specific topic and photo/figure numbers. In some instances photo/image numbers may not be sequential or may be absent. This is intentional.

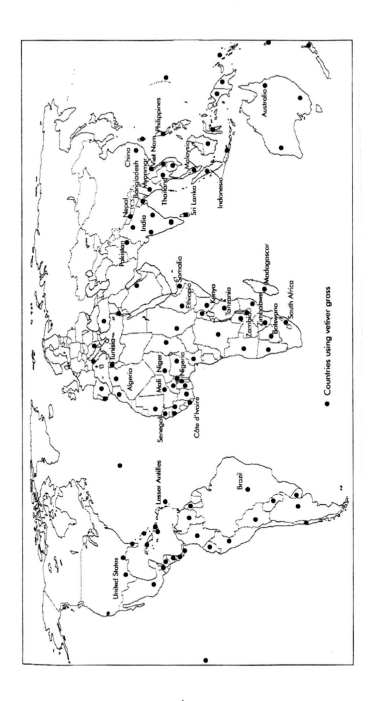

The Vetiver System
for
Soil and Water Conservation

Proven and Green Environmental Solutions

Sheet Erosion

Under normal conditions, sheet erosion is not recognized and therefore is seldom treated. However, triggered by torrential rainfall, sheet erosion accounts for the loss of thousands of lives through mud slides and landslides, and the loss of billions of tons of soil every year. As raindrops pound the ground, particles of soil are knocked loose and carried away by the water runoff. This runoff further strips unprotected areas of their valuable topsoil and becomes the muddy water that ends up in drains, streams, and rivers. Sheet erosion leads to more striking forms of erosion - rills and gullies, for example, the focus of most conservation efforts to date.

Although not as spectacular as rills and gullies, sheet erosion does leave visible marks, as shown in Figure 1: soil collecting behind obstructions on a slope (such as the brick in example A); stones left behind by the runoff because they were too heavy to be carried away (B); and molded forms on the of soil and other debris trapped under branches, twigs, or even clumps of straw (C). The reality of sheet erosion's effects appears in Photo 1.

The effects of sheet erosion are more readily apparent in forest areas that are devoid of ground cover, and in fields or wastelands with a few standing trees, where the loss of soil exposes the roots of the trees. (See Figure 2 and Photo 2.) Water can then easily pass beneath the trunks of the trees and among their roots. After all, if the soil that supported them and gave them life is washed away, the trees will be washed out of the ground as well.

Figure 1. Sheet erosion leaves visible marks.

Photo 1. Land shows sign of sheet erosion.

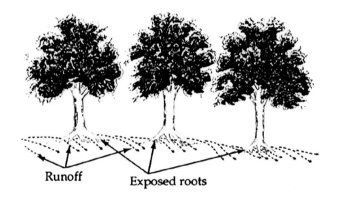

Figure 2. Sheet erosion exposes tree roots.

Photo 2. Loss of soil exposes roots.

Trees by themselves do not prevent soil loss caused by sheet erosion; forests do, with their thick litter and low-growing vegetation. In areas where forest cover is not possible or practicable, vegetative barriers can be used to stop the loss of soil. Fibrous-rooted shrubs and grasses planted as hedges along the contour or across the slope of the land slow the runoff, spread the water about, weaken its erosive power, and cause it to deposit its load of valuable soil behind the hedgerows. As a result, the runoff proceeds gently down the slope, and if the hedges have been planted at the correct vertical interval (discussed later and illustrated in Figure 23), it proceeds without further erosive effect.

The amount of soil lost through sheet erosion is alarming. Figure 3, depicts two surviving plants whose roots prevent sheet erosion, and shows how the amount can be measured. In this case a layer of soil 50 centimeters deep - as measured by the distance between the top of the plant mounds and the present soil surface - has been lost across the entire area of the field since the plants became established. Photo 3 shows the vulnerability of the exposed roots after the soil has washed away.

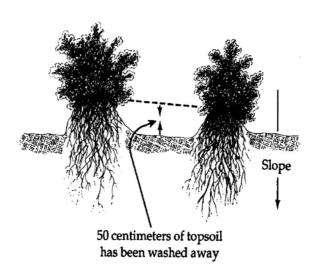

50 centimeters of topsoil has been washed away

Figure 3. Sheet erosion washes away topsoil.

Photo 3. Sheet erosion results in loss of topsoil.

Rainfed Farming

The traditional way of farming in rainfed areas, no matter how flat the land may seem, for easy application, is along the slope, or up and down the hill. (Figure 4 and Photo 4). This system encourages runoff and soil loss and thus makes sheet erosion worse. Often more than

Figure 4. Traditionally farmers in rainfed areas tend to plant along the slope or up and down the hill, thus reducing moisture conservation and increasing run off and soil loss.

Photo 4. Most of these Venezuelan farmers are cultivating up and down the slope resulting in very large amounts of soil loss and runoff.

50 percent of the rainfall is lost as runoff, and is thus denied to the crops; and the steeper the slope, the faster and more erosive the runoff. Rainfall is less effective because the water is not given a chance to soak in to the soil. By plowing along the slope, the farmer in Figure 4 is unknowingly encouraging the loss of precious rainfall.

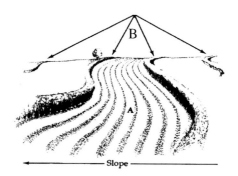

Figure 5. Farming with vegetative contour hedges reduces erosion and conserves rainfall for better plant growth and groundwater recharge.

Figure 5 and Photo 5 illustrate the method advocated in this handbook - the use of vegetative contour hedges to prevent erosion and conserve

Photo 5. Farmers who use vegetative contour hedges, that require little or no maintenance, will protect their fields from erosion for many years, and will conserve moisture for sustained cropping.

natural moisture in the soil. Once established, such hedges need little maintenance and will protect the land from erosion for years as they build up (particularly on steep slopes) natural terraces. In contrast to the planting furrows in Figure 4, those at A in Figure 5 follow the contour of the land as laid out by the vegetative hedges at B in the illustration.

Constructed earthen embankments, or contour bunds, have slowed erosion throughout the world since the 1930s. But this constructed method of soil conservation creates an unnatural system of drainage that has diverted unnecessarily large quantities of water from crop land, and, from my experience in the field, I no longer consider it an appropriate soil conservation practice, especially for small farmers in tropics.

The embankment in Figure 6 was constructed with topsoil taken from point A, and transformed into a channel to convey the runoff sideways. But the bank is made of the same soil it is supposed to protect, and because its construction makes the slope steeper, over time the bank will erode and "melt" away. Then it will have to be replaced - at great cost to the farmer. Moreover, to collect sufficient soil to make the bank and channel shown in Figure 6, a 5-meter-wide strip of land must be taken out of production over the entire length of the bank. This represents a loss of 1 hectare of productive farmland for every 20 hectares of land treated with embankments or bunds. Photo 6 clearly shows a typical failure of the constructed system of conservation in self-mulching vertisols in India.

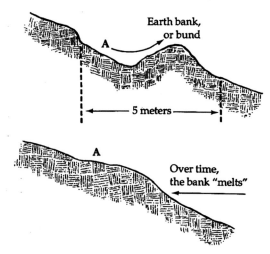

Figure 6. Constructed soil conservation works need considerable maintenance, involving much labor and cost, and over time will erode away.

Photo 6. With the constructed method of soil conservation, conservation efforts failed on India's black cotton soil.

Figure 7 shows the unnatural way the land is drained by the constructed system. All of the runoff is channeled sideways and dumped into a constructed but unproductive waterway that no smallholder would want running through his or her farm. This system makes the areas below the banks too dry and the channel areas too wet for optimum

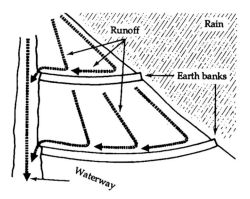

Figure 7. Constructed soil conservation works take up a 5-meter-wide strip of land over the entire length of the bank, and deflects precious rain water off the land into waterways that often turn into gullies.

crop production. In contrast, the vegetative method of soil and moisture conservation uses nature to protect itself. In the system demonstrated in this handbook with vetiver grass, only a 50-centimeter strip - or one-tenth of the land occupied by earthen embankments or bunds - soil is disturbed; whereas earth banks have to be made with bulldozers or by hired labor. The vegetative system requires no special tools or labor beyond that a farmer would already have.

Photo 8. Over a thirty year period, a two meter terrace riser built up naturally behind this vetiver hedgerow in Fiji. This represents a huge amount of soil retained on the farm.

The bottom illustration in Figure 8 shows what happens over time in the vegetative system: the runoff drops its load of soil behind the vetiver hedge, the grass tillers up through this silt, and a natural terrace is created. The terrace becomes a permanent feature of the landscape, a protective barrier that will remain effective for decades, even centuries. Photo 8 demonstrates how the terrace builds up over time. Also Photo 41 page 63 shows a Google Earth image of 50 year old terraces in Fiji created by vetiver hedgerows.

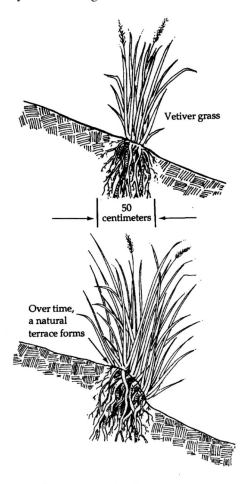

Figure 8. The vegetative system of soil and moisture conservation uses nature to protect the soil.

Photo 9a Not only has the constructed bank failed, but the cotton crop that the bank was constructed to protect has also failed. Before this bank's failure, the constructed system trapped too much water, flooding the entire area. Because cotton cannot tolerate poor drainage, the crop failed completely.

Photo 9b. This field is in the same area as that shown in Photo 9, but here the cotton crop is protected by a single vetiver hedge. The hedge spreads out the runoff, filtering it and giving it a chance to soak into the ground, and should result in an excellent crop of cotton.

Photo 9a shows not only the failure of a constructed bank in India but also the failure of the cotton crop the bank was designed to protect. Before this bank's failure, the entire area was flooded by the constructed system's trapping of too much water. Because cotton cannot tolerate poor drainage, the crop failed completely.

Photo 9b was taken on the same day in a nearby location to Photo 9a. In this situation the cotton crop protected by a single vetiver hedge that spread out the runoff and gave the water a chance to soak into the ground over the whole area protected by the hedge; thus producing a potentially good crop of cotton. With the vegetative system, when the runoff reaches the hedges, it slows down, spreads out, drops its silt load, and oozes through the hedgerows, a large portion of the water soaking into the land along the way (Figure 10.). No soil is lost, and there is no loss of water through the concentration of runoff in particular areas. The Vetiver System requires no engineering—the farmers, with a little training, can do the whole job themselves.

Near Mysore, in the southern Indian state of Karnataka (in the villages and hamlets of Gundalpet and Nanjangud, for example), farmers have been maintaining vetiver hedges as boundary markers around their farms for more than 100 years. To keep the hedges narrow, the farmers simply plow around the edges of the hedgerows whenever they plow the rest of the field for cropping. The hedges are in perfect condition and provide permanent protection against erosion (Photo 10).

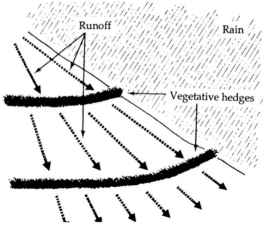

Figure 10. With the vegetative system of soil and moisture conservation, filtering benefits the crop. Runoff slows down, spreads out, drops its silt load, and oozes, or filters through the hedgerows, enabling a large portion of the water to soak into the land along the way.

Photo 10. When the runoff drops its silt load, it builds up behind the hedge, creating a natural terrace. The height of the terrace is demonstrated here by the line of farmers; the one on the right appears taller because he is standing on the terrace. This hedge on gently sloping farm land in India may be 100 years old. Note this hedgerow is a perfect example of a continuous hedge filter barrier with no breaks or gaps. This guarantees success and is one reason that runoff in the area is significantly reduced resulting in improved ground water levels. The hedge is cut regularly for livestock forage.

Vegetative Contour Hedges

Figure 11 presents a cross-sectional view of a vetiver contour hedge at work. The leaves and stems of the vetiver plant slow the silt-loaded runoff at A and cause it to deposit the silt behind the plant at B while the water continues down the slope at C at a much slower pace. The plant's spongy root system, pictured at D, binds the soil beneath the plant to a depth of up to 3 meters. By forming a dense underground curtain that follows the contour of the land, the roots prevent rilling, gullying, and tunneling. The photo 11 shows how the soil from the runoff has been trapped above the original topsoil behind the vetiver hedge.

Figure 11. This cross-sectional view of vetiver shows a vegetative contour hedge at work.

Photo 11. This cross-sectional view of a two-year-old vetiver hedge in Malaysia, shows how about 40 cm of soil has been trapped by the hedge above the original dark band of topsoil (P.K.Yoon).

The strong aromatic oil contained in the vetiver roots makes the grass unpalatable to rodents and other pests, and many Indian farmers report that it also keeps rats from nesting in the area. Because the dense root system repels rhizomes of grasses such as *Cynodon dactylon*, the hedgerows prevent such grasses from entering the farm field and becoming a weed. Another benefit of planting the hedgerows, according to the farmers near Mysore, is that the plant's sharp, stiff leaves keep away snakes (this may be because there are no rats).

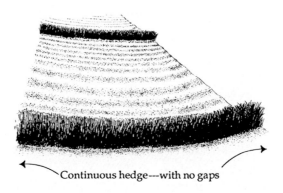

Figure12. With the vegetative system, the hedge grows continuously, leaving no gaps for runoff to cause rilling or gullying.

Photo 12a. A vegetative system in Panama conserves soil and moisture, protecting the crop

To be effective as a method of soil conservation, the vegetative system must form a hedge, as shown in Figure 12 and in Photos 12a and 12b.

Photo 12b. In Costa Rica, a vetiver hedge hugs the contour of the land, conserving soil and moisture and protecting the crop.

Although under certain circumstances thick hedges can be formed in one year, it generally takes two to three growing seasons to establish a hedge dense enough to withstand torrential rains and protect the soil. During the first two seasons, and sometimes the third, the vetiver plants need protection, and any gaps in their line have to be filled. (During the first two seasons it should also be easy to see the silt being trapped behind the plants as they are establishing, a phenomenon that extension workers should try to point out when explaining the system to farmers.) Although the earth banks used in the constructed method of soil conservation are effective immediately, they break down over time and frequently burst open in heavy rainstorms. Once a vegetative hedge has been established, it will neither wear out nor require further maintenance, other than periodic trimming. The photos in Figures 12a and 12b provide a good example of this use of the Vetiver System.

Trimming the hedges to a height of 30 to 50 centimeters prevents them from flowering and makes them thicken up, thereby increasing their effectiveness in filtering runoff and improving groundwater recharge. In several villages and hamlets near Mysore, the farmers trim their hedges every 3 weeks throughout the year and feed the young palatable leaves to their livestock, often chopping them up and mixing them with other fodder. They are thus ensured a year-round supply of stock fodder regardless of rainfall.

Following the Contour

Many field workers - and even research workers - lack a clear understanding of what is meant by the *contour*. Figure 13a illustrates a common misconception: that a furrow plowed along the main slope follows the contour. This is incorrect.

A true contour embraces all slopes, major or minor; it is a line of equal elevation around a hill. The furrows in Figure 13a, which starting from point A, follow the main slope straight down to point C, instead of curving around the hill; they are not on the contour and therefore will neither conserve moisture nor prevent erosion. The true contour pictured in Figure 13b, runs from A to B to D and continues around the hill, maintaining equal elevation all the way.

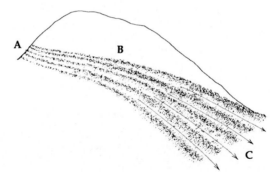

Figure 13a. With this false contour of the land, the furrows follow the main slope straight down instead of curving around the hill. This contour will not conserve soil or moisture.

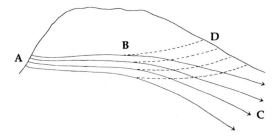

Figure 13b. With this true contour of the land, the furrows embrace all slopes and maintain equal elevation all the way. This contour will conserve soil and moisture.

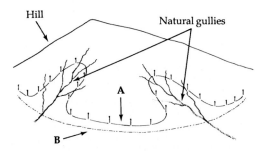

Figure 13c. Earth banks, that conveys runoff to a waterway off to the side of the field, must be constructed on the exact contour (marked with pegs at A), often making it difficult for the farmer to follow when plowing. With vegetative hedges, however, the contour can be averaged into a smooth curve (line B) and the hedge planted across the slope.

Constructed earth banks that conventionally are used to control erosion must convey the runoff to a waterway off to the side of the field, and therefore they have to be constructed on the exact contour as shown in Figure 13c; such a line (marked with pegs at A) can be difficult for the farmer to follow when plowing.

The vetiver hedges, however, do not have to be exactly on the contour to provide effective soil and moisture conservation since their purpose is to reduce the velocity of the water as it passes through them and not to channel the water elsewhere. After the contour line has been pegged in (see Figure 18a, page 26), the extension worker can smooth it out to make it easier for the farmer to follow. In Figure 13c hedges and plow furrows (crop lines) need only follow line B. The silt filtered from the runoff will build up behind the hedges and eventually form a natural terrace. Because the hedges run across the slope, the ends of each hedgerow should be turned up the slope to prevent runoff from spilling around the sides; this will encourage natural terraces to form more readily and prevent erosion at the ends of the hedgerows, especially in steep lands.

In Figures 14a and 14b and Photo 14 we have two farmers, A and B. Both are esperienced farmers, but farmer A in Figure 14a is more sensible and has protected his land against soil loss by planting vetiver hedges on the contour, and he is using the hedgerows as guidelines to plow and plant on the contour.

Figure 14a. On a protected farm, vegetative hedges planted on the contour protect this land against soil loss. The hedgerows serve as guidelines for plowing and planting on the contour, creating furrows that will hold rainfall and store extra moisture in the soil in anticipation of long periods of dry weather.

The furrows created in this fashion will hold rainfall and store extra moisture in the soil, thus allowing crops to withstand long periods of dry weather. What farmer A is doing costs no more than what farmer B in Figure 14b is doing. All that is involved is a change in practice.

Photo 14. Vegetative hedges planted on the contour protect this land against soil loss. Also the hedgerows serve as guidelines for plowing and planting on the contour, creating furrows that will hold back rainfall and store extra moisture in the soil in anticipation of long periods of dry weather.

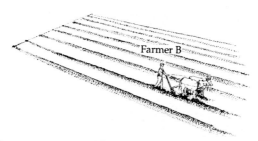

Figure 14b. On an unprotected farm, plowed furrows that run straight up and down the slope encourage the rainfall to run off the farm, taking soil and farmyard manure for the ride and moving so quickly that no water is soaked into the soil as a protection against dry spells.

Farmer B is not farming wisely. By plowing just straight up and down the slope, even a very gentle slope, he is encouraging the rainfall to run off his farm, taking his farmyard manure (FYM) and an irreplaceable layer of topsoil along with it.. The rainwater runs off so quickly it does not have a chance to soak into the soil, and thus his crops have no protection against dry spells, and as a result under drought conditions most annual farm crops will wilt more quickly.

Figure 15a. When rain falls on the protected farm, the vegetative hedges and the contour furrows protect the soil from running off the land.

Figures 15a and 15b and Photo 15a illustrate what happens when the two farming systems are exposed to heavy rainfall. Farmer A's field is protected by the vegetative hedges, and there is no loss of soil (Figure 15a and Photo 15a). The contour furrows store all the rainwater they can hold. Any surplus rainfall runs off, but the vetiver hedges control the flow — slowing it down, spreading the water out — and causing the silt to be deposited. As a result, the runoff is conducted down the slope in a safe, non-erosive manner.

Photo 15a. When rain falls on the protected farm, the vegetative hedges and the contour furrows protect the soil from running off the land. In this photo on India's black soils you can see a clear band of crop residue spread behind the hedgerow due to the even and temporary back up of water along the length of the hedgerow.

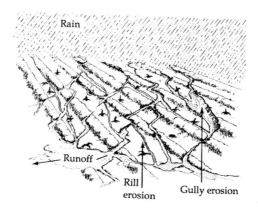

Figure 15b. When rain falls on the unprotected farm, the water runs off at great speed, taking the topsoil and fertilizers with it and eroding the soil as it moves along.

On Farmer B's unprotected land, the rainfall runs off at great speed, taking along his fertilizers and topsoil. The uncontrolled flow down the slope causes unnecessary and damaging erosion (Figure 15b). Because the runoff is so fast, no moisture is stored. Rainfall is only 40 to 50 percent effective, and farmer B is always complaining about droughts. Ultimately he will have to abandon his farm because there will be no soil left in which to grow crops. Farmer A will never have this problem; his yields will increase over the years.

Photos 15b and 15c show the importance of land management for moisture conservation in rainfed areas even on very flat land. Unlike irrigated farming, in which farmers have complete control over their crops' water needs, rainfed farming is totally limited by the amount of rain that falls in the area for the success or failure of the crops. By plowing and planting on the average contour, or across the slope, rainfed farmers have a better chance of holding the rain that falls in the field and in the actual crop rows, which means that the whole field benefits. In heavy storms, when the crop rows cannot hold the rainfall, the vetiver hedges prevent damage from erosion by spreading the runoff, taking the power out of it, and giving it time to soak into the ground. Thanks to his vetiver contour hedges, farmer A obtains an excellent crop (Figure 16a and Photo 16). Because the soil has retained ample moisture the from earlier rains, his crop is benefiting from the

Photo 15b. Because rainfed farmers are limited to the amount of rain that falls in their area for the success of their crops, they must plow and plant on the average contour to hold the rain that falls on the field and in the actual crop rows. The farmers also need vetiver hedges to prevent an abundance of rainfall from eroding the soil by spreading the runoff and giving it time to soak into the ground. On nearly flat lands, as in this photo of cropland in Australia, when flooded, vetiver hedges significantly reduce water velocity and related crop damage.

Photo 15c. When farmers do not plow and plant on the average contour, heavy rainfall events will wash away the topsoil and fertilizers. Without vetiver hedges to spread the rainfall runoff, giving it time to soak into the soil, large quantities of precious rainfall will be lost.

warm sunshine, all grains are filling, and the crop stand shows even growth. Farmer A will reap a high yield. In contrast, farmer B has a disappointing harvest. (See Figure 16b.) His crop has all but failed, and what little remains - growing in pockets where some moisture was trapped - is being dried out by the sun. Only a small percentage of the grain will fill, and the resulting crop is uneven.

Figure 16a. (Left) On the farm with a protective vetiver grass contour hedge, farmer A obtains an abundant crop. Figure 16b. (Right) On the unprotected farm, farmer B's failing crop grows only in pockets where some moisture was trapped; but the sun will soon dry out the crop.

Photo 16. This is how an Ethiopian farmer followed Farmer A's approach. An excellent maize crop planted on the contour, protected by a vetiver hedgerow. This hedgerow also acts as a habitat for beneficial insects, and is the preferred host for maize stem borer, (the stem borer moth lays its eggs on the vetiver leaves, the larvae hatch, fall off and die thereby reducing the incidence of stem borer on the maize crop by as much as 90%). Thus the hedgerow has an integrated pest management role as well.

Photo 17a. A young vetiver hedgerow protecting a maize crop in Malawi.

Farmer B can expect a low yield. Yet he planted the same crop as farmer A, used the same fertilizer, planted at the same time, and received the same amounts of rainfall and sunshine. Unlike his neighbor, however, farmer B lost most of his fertilizer, 60 percent of his rainfall, and a layer of soil, possibly a centimeter thick, from his farm - all because he did not plow on the contour and use vegetative hedges to protect against erosion and help his cropland retain moisture from the rain.

If he had taken the advice of his extension service and plowed and planted on the contour, farmer B could have obtained the same high yields as farmer A. Photos 17a and 17b attest to the success of crops that are protected by vetiver hedges. Having learned his lesson, farmer B contacts his extension worker, and together they mark, or peg out, contour lines across the old furrows. (See Figure 18a.) This simple process requires virtually no engineering skills - only the use of a small hand-held level (or A frame). The extension worker stands at the edge of the field and, sighting through the level, has farmer B move up or down the slope until the two people are standing level, at which point the farmer marks the spot with a peg. In Figure 18a, the contour line (X) has already been pegged out, and the farmer has but to follow the line of pegs with his plow (as shown in Figure 18b) to create the furrow in which to plant the slips of vetiver grass that will eventually

Photo 17b. Vetiver hedgerows act as a windbreak on Pintang Island, China, protecting jojoba seedlings from soil and wind erosion.

form a contour hedge. This is all that has to be done to establish the vegetative system of soil and moisture conservation. Photo 18 shows an extension worker and farmers working together to peg out vetiver hedges on the contour.

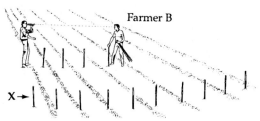

Figure 18a. Farmer B and his extension worker peg out contour lines along the old furrow.

Figure 18b. Farmer B plants on the contour. Hedges are just as effective planted across the slope, straight across rills and small gullies. Planting across the slope makes it easier for the farmer to cultivate as there are no sharp turns.

Like any long-lived plant, however, the vetiver hedge system normally takes two to three seasons to become fully effective. You cannot plant a mango tree today and expect to pick mangoes next month, but it is possible to get some immediate effect from the system by using dead furrows as a preliminary step until such time as the vetiver grass can be established.

Photo 18. An extension worker in Papua New Guinea pegs out an average contour line for a vetiver hedgerow.

The preliminary stage of establishing the vegetative system is depicted in Figure 19a. While waiting for vetiver planting material to be produced in the nursery, the farmer laid out the contours, prepared seedbeds following the contour furrows, and every 5 or 6 meters double plowed a dead furrow. The two dead furrows in the figure have been planted on the contour to pigeon peas and inter cropped with six rows of groundnuts. The shape of each seedbed is shown beneath the crop illustration: DF marks the deeper dead furrow, PP the row of pigeon peas it supports. Eventually, vetiver grass will be planted in some of the dead furrows, but in the interim these furrows themselves will provide a bit of protection against runoff. Planting the vetiver grass will stabilize the whole system, as shown in Figure 19b, where a vetiver hedge has taken the place of one of the dead furrows.

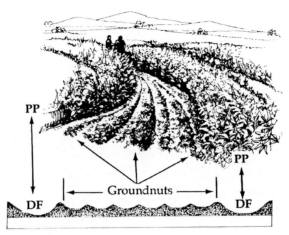

Figure 19a. The initial setup of vetiver hedges uses dead furrows (DF) to support the pigeon peas (PP).

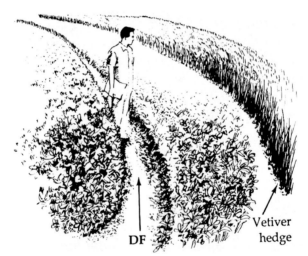

Figure 19b. To stabilize the system, the farmer has planted a vetiver hedge in one of the dead furrows.

Establishing Vetiver Hedges

To establish a vetiver hedge, follow the step-by-step instructions that appear on the next few pages, along with tips on handling the planting material, advice about the best time to plant, and information about what to expect after the grass is planted. The first step for establishing a vetiver hedge is obtaining the planting material, usually from a vetiver nursery. If vetiver grass is unknown in your area, check with the nearby botanical gardens. Ask them to look up *Chrysopogon zizanioides*. If it has been collected, the herbarium sheet will show what the plant looks like, note where the specimen was found, and provide the local name of the plant. Vetiver is found throughout the tropics and has been grown successfully as far north as 42° latitude. Vetiver nurseries are easy to establish. Inlets to small dams or water holding tanks make the best nursery sites, because water en route to the dam or tank irrigates the vetiver grass, which in turn removes silt from the water. Large gullies protected with vetiver grass also make good informal nurseries. For best results in establishing a vetiver nursery, the vetiver root divisions, or slips, should be planted in a double or triple line to form parallel hedges across the streambed. The hedgerows should be about 30 to 40 centimeters apart.

To remove a clump of vetiver grass from the nursery, as shown in Figure 20 (A), dig it out with a spade or fork. The root system is too massive and strong for the grass to be pulled out by hand. Next, tear a handful of the grass, roots and all, from the clump (B). The resulting piece, the slip, is what will be planted in the field (C). In Photo 20, a Cambodian farmer has dug up a good clump of vetiver - one doesn't need much root. Before transporting the slips from the nursery to the field, cut the tops off about 15 to 20 centimeters above the base, and the roots 10 centimeters below the base (the old roots do not regrow but act as an anchor). Cutting will improve the slips' chances of survival after planting by reducing the transpiration level and thereby preventing them from drying out. As shown in Figure 21a, all that is needed to prepare the slips for planting is a block of wood and a knife—a cane knife, machete, cutlass, or panga will do.

Figure 20. The farmer digs out a clump of vetiver grass from the nursery (A), tears a handful of grass and roots (the slip) from the clump (B), and prepares to plant the slip in the field (C). A minimum of three tillers per slip seems to work best.

Photo 20. This Cambodian farmer, living adjacent to the Mekong River, uses a narrow bladed spade, rather like a oil palm fruit harvesting knife (the latter is probably the best tool for the job) to dig up a plant from a quality nursery (Tuon Van).

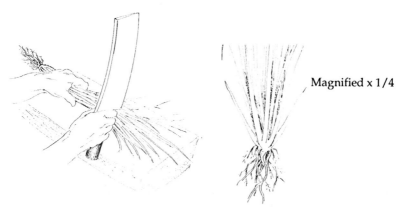

Figure 21a. (left): To prepare a vetiver slip for planting, the farmer holds the slip on a block of wood and uses a knife to trim the roots and the leaves; **(right):** The trimmed vetiver slip are ready for planting.

Photo 21a. The farmer trims the slips. Nice job, good quality tillers (3 tillers per slip).

Although vetiver grass can be planted from single tillers (when planting material is scarce), this practice is not recommended for grass to be planted in the field, because it takes too long to form a hedge. Fertilizing the slips with di-ammonium phosphate (DAP) encourages fast tillering and is helpful both in the nursery and in the field. To do this in the field, simply dibble DAP into the planting furrow before planting the slips.

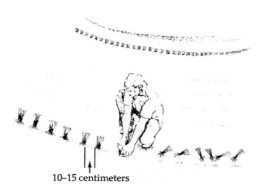

Figure 21b. The farmer plants the vetiver slips 10-15 centimeters apart.

Photo 21b In Mexico, a farmer has correctly spaced planting slips in a vetiver hedgerow.

Always plant the slips at the beginning of the wet season to ensure that they get full benefit of the rains. Planting vetiver slips is similar to planting rice seedlings. Make a hole in the furrow that was plowed to mark the contour. Push the slip into the hole, taking care not to bend the roots upward, firm the slip in the soil, and then 10 to 15 centimeters from the slip, along the same contour furrow, plant the next slip, and so on. (Figure 21b and Photo 21b)

Only a single row of slips need be planted. If planted correctly, the slips can withstand up to one month of dry weather. Some slips may die, however, and leave gaps in the hedge line. If possible, fill these gaps by planting new slips. In some instances it may be possible to use the live flower stems, or culms, of neighboring plants—simply bend the culms over to the gap and bury them. The live stems will produce roots and leaves at the nodes.

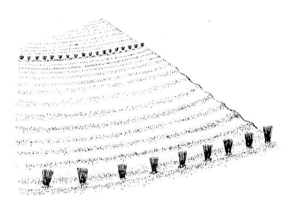

Figure 22. If the farmer plants the slips too far apart, it will take too long for the hedgerow to grow.

Photo 22. In China, a farmer incorrectly spaced the planting slips and the vetiver hedgerow never formed. Without a dense hedge the system will not work.

Of course for this or any vegetative system to work, the plant must form a hedge; otherwise, the system cannot act as a barrier against soil loss nor can it effectively reduce runoff. Planting the slips too far apart (see Figure 22) would render the system almost useless because it would take too long for the slips to grow together to form a hedge. Photo 22 shows a hedge in China that was planted the wrong way. Even though the farmer had the extension workers' handbook showing the correct method for achieving a hedge with vetiver and had advice in the field, he planted the slips too far apart. This method of planting will never work as a method of soil conservation, and the farmer will eventually abandon the system - not because the system does not work, but because the farmer did not lay it out according the instructions.

Moreover, without the extra support of a hedge to hold the soil, fertilizer, and moisture against the vetiver grass, the plants would not be able to survive the worst droughts. Even in arid areas that receive less than 200 millimeters of rain a year, an effective vetiver contour hedge could ensure its own viability. The combined effect of contour cultivation and the hedge's performance in slowing and spreading the runoff is to increase infiltration of water into the soil. Thus the hedge can help itself to what might be the equivalent of half again as much rainfall.

For the system to provide maximum protection against erosion, the hedgerows should be spaced apart at the proper vertical interval (VI). The VI is the vertical distance from one hedgerow to the next one down the slope. The actual distance measured along the ground, called the surface run, depends on the steepness of the slope. With a vertical interval of 2 meters, for example, the hedges on a 5 percent slope would be about 40 meters apart, whereas those on a 2 percent slope would be about 100 meters apart. As shown in Figure 23, the surface run between hedgerows planted on a 57 percent slope with a VI of 2 meters is about 4 meters. For a more comprehensive look at the relationships among slope, surface run, and vertical interval, see Table 1 at the end of this handbook. In practice, a VI of 2 meters has generally been found to be adequate.

After the hedges have been established in the farm field, the only care they will need is annual trimming to a height of about 30 to 50 centimeters to encourage tillering and prevent shading of the food crops. When plowing for cropping, plowing along the edges of the hedgerows will remove any tillers that encroach upon the field and will thus prevent the hedges from getting too wide.

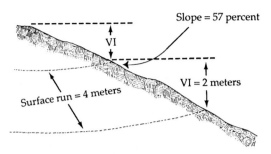

Figure 23. The surface run between hedgerows planted on a 57 percent slope with a vertical interval of 2 meters is about 4 meters.

Moisture Conservation

Although measures to retain natural moisture in the soil are essential to all rainfed farming systems, the art of in situ moisture conservation, as it is called, is rarely practiced and not widely understood. There is no such thing as flat land; water runs off all land. No matter how flat it may seem, all land must be protected if it is rainfed. Earth shaping, land leveling, and similar techniques are required in irrigated areas only; rainfed areas must be contoured. Figure 24 shows what happens when land is planted on the "flat" without the benefit of contour furrows.

In view A, the rain runs straight off the field. View B shows the results: because no moisture has been stored, the plants wilt and die in the sun. View C shows the same area planted to contour furrows, with a pair of dead furrows taking up the surplus runoff until the vetiver can be planted. Rain caught and held in each furrow's micro-catchment has the chance to soak in. Each furrow can hold 50 millimeters of rainfall, so in most storms there is no runoff. Thanks to this natural system of water storage, the plants can benefit from the sunshine, as shown in view D. In view E, one of the dead furrows has been planted to vetiver grass to stabilize the system.

A vetiver grass hedge is the key to the in situ moisture conservation system. Once established, it serves as a guideline for plowing and planting on the contour, and in times of heavy storms it prevents erosion from destroying the farmer's field. The beauty of the plant is that, once it has established, the hedge is permanent.

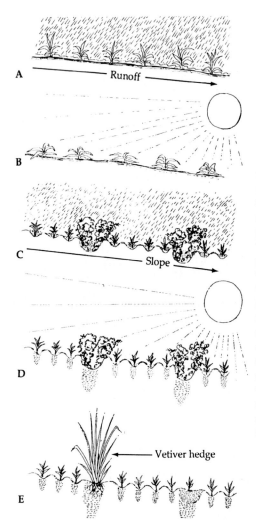

Figure 24. When land is planted on the flat, rain runs straight off the field (A), resulting in land with no stored moisture and plants that wilt and die in the sun (B). When land is planted on the contour, while waiting for the time to plant vetiver, dead furrows take up the surplus runoff (C), resulting in a natural system of water storage and plants that enjoy both the moisture and the sun (D). When vetiver grass is planted in one of the dead furrows, it stabilizes the natural system (E).

Figure 25 is a diagrammatic representation of what a vetiver grass system would look like in a smallholder farming area. The Vetiver System fits perfectly into the individual farm system, where no waterways or earthworks exist. Most farmers have one line of vetiver

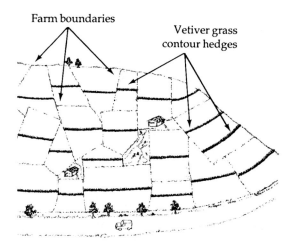

Figure 25. This is what a system of vetiver hedges planted over a large area would look like. Unlike a constructed system, the Vetiver System requires no waterways and does not have to spill into a drainage network.

Photo 25. The Vetiver System (Indonesia) enables the farmer to plant vetiver the full length of his field since the hedges do not convey runoff.

roughly in the middle of their fields, no matter what the shape; long fields may need two lines to stabilize them. Although each field has its own line or lines of vetiver, the entire hillside is protected against erosion because each line protects the ones farther down the slope.

Under this system, once the hedges are established, no further protective work is needed, and maintenance is minimal. The farmers each have their own supply of vetiver planting material. Should a gully start to form anywhere, the farmer obtains slips of vetiver from an existing hedge and plants it across the incipient gully to prevent its spread—permanently and at no cost except for the farmer's own labor.

Photo 25 shows vetiver hedges planted over a large area. Unlike a constructed system, the Vetiver System requires no waterways, and runoff does not have to spill into any drainage network. Because they do not convey runoff, the hedges can be as long as any given farmer's field. The hedges spread out the runoff, providing moisture conservation over the entire area. At the same time, the hedges filter out the silt, enabling the excess runoff to flow harmlessly down the slope.

Why Vetiver Grass Is the Ideal Plant for the Vegetative System of Soil and Moisture Conservation

Although many grasses and trees have been tried over the years as measures to prevent erosion, to date only vetiver grass has stood the test of time. As made clear by the following list of its characteristics - derived from observations of *C. zizanioides* throughout the world - this truly remarkable plant is ideally suited for the vegetative system of soil and moisture conservation. No other grass is known to rival its hardiness or diversity.

C. zizanioides:
- When planted correctly (i.e. close together), it will quickly form a dense, permanent hedge.
- Has a strong fibrous root system that penetrates the soil vertically and binds the soil to a depth of up to 3 meters and can withstand the effects of tunneling and cracking.
- Is perennial and requires minimal maintenance.
- Is practically sterile - it produces no stolons or rhizomes, it seeds are sterile (south Indian cultivars). It will not become a weed.
- Has a crown that is below the surface, which protects the plant against fire and overgrazing.

- Has sharp leaves and aromatic roots that repel rodents, snakes, and similar pests.
- Has leaves and roots that have demonstrated a resistance to most diseases.
- If managed correctly it makes an excellent fodder. The young leaves are palatable and high in protein (In Karnataka, India, a cultivar of *C. zizanioides* selected by farmers has softer leaves and is more palatable to livestock. This cultivar is also denser, less woody, and more resistant to drought than some of the other available cultivars).
- Is both a xerophyte and a hydrophyte, and once established, vetiver grass can withstand drought, flood, and long periods of waterlogging.
- Will not compete with the crop plants it is used to protect, and in fact, vetiver grass hedges have been shown to have no negative effect on - and may often boost - the yield of neighboring food crops.
- It has associated nitrogen processing mycorrhiza, that explains its green growth throughout the year.
- Is cheap and easy to establish as a hedge and to maintain—as well as to remove if it is no longer wanted.
- Will grow in all types of soil, regardless of fertility, pH, or salinity, including sands, shales, gravels, and even soils with aluminum toxicity.
- Will grow in a wide range of climates and is known to grow in areas with average annual rainfall between 200 and 6,000 millimeters and with temperatures ranging from a 'flash frost" of -14° to +55° Centigrade.
- Is a climax plant; therefore, even when drought, flood, pests, disease, fire, or other adversity destroy all surrounding plants, the vetiver will remain to protect the ground from the onslaught of the next rains.

NOTE: THESE CHARACTERISTICS APPLY TO THE NON FERTILE CULTIVARS OF C. ZIZANIOIDES ORIGINATING FROM SOUTH INDIA.

Other Practical Uses for Vetiver Grass

Apart from its success as a system of soil and moisture conservation, vetiver grass has proved effective for a variety of other purposes.

Stabilizing terrain and structures
One of the most important uses of vetiver grass is stabilizing the terrain and structures such as dams, canals, and roadways.

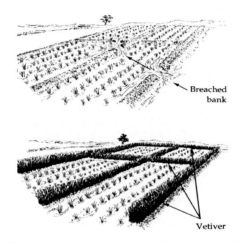

Figure 26. Vetiver can be used to stabilize a paddy field banks.

Figure 26, for example, shows how vetiver can be used to stabilize a typical paddy field that relies on earth banks to keep irrigation water at the correct level. These banks can be worn down by the action of wind-churned water (lap erosion) and the activities of rats, crabs, and other hole-burrowing pests. The subsequent large-scale erosion, not to mention the loss of expensive and in some cases irreplaceable irrigation water, could lead to loss of the crop.

Vetiver can be planted on top of the paddy banks to stabilize them. Vetiver grows well under these conditions and does not suffer from the occasional inundation of water. In addition, vetiver roots contain an essential oil that repels rodents. Furthermore, because its roots grow straight down and not out into the crop, the grass has no negative effect on the rice or its yield.

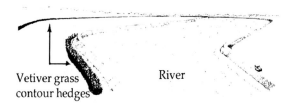

Figure 27. Vetiver protects farmland by stabilizing river banks and by preventing river levees from breaching and flooding into the fields.

Photo 27. In Zimbabwe, vetiver protects a riverbank from a fast flow river.

Each year the vetiver can be cut back to ground level to prevent shading of the crop and be fed as forage. In another example, vetiver can be used to maintain river levees and will prevent serious levee breaching under flood conditions - a practice now being followed in coastal Vietnam. Vetiver can also be used on river banks to prevent silt from entering the watercourse from the runoff of surrounding fields, as well as preventing river bank erosion (Figure 27 and Photo 27).

Establishing tree crops

Vetiver's stabilizing influence is especially useful in steep and rolling country, where the distribution of moisture cannot be controlled. Unsuitable for the cultivation of cereal or other annual crops, such areas, when stabilized by vetiver grass, can be successfully planted to perennial tree crops on the contour. Most attempts to grow tree crops on steep hillsides are abandoned because the resulting poor, uneven stands are not worth the cost of maintenance.

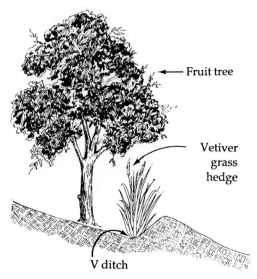

Figure 28. A vetiver grass hedge nurtures a fruit tree on a hill.

Figures 28 and 29 show a method of establishing tree crops on such hills using contour vetiver hedges. First the contours of the hill are pegged out. Next, by hand or with a bulldozer and ripper unit, the farmer digs shallow V ditches along the contour lines. A row of trees is planted close to the edge of each ditch, and vetiver grass is planted in the ditches.

Under this arrangement of planting, the runoff between one row of trees and the next one down the slope collects in the vetiver-lined ditches (there is usually sufficient drainage on the slopes to preclude the possibility of waterlogging). Thanks to the effects of such water harvesting, the rows of trees do not have to be planted as close together as the trees within a row. Initially, the V ditch will provide a measure of runoff control, thereby increasing the soil's moisture content,

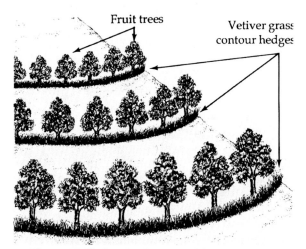

Figure 29. Vetiver hedges stabilize tree crops.

and both the vetiver and the planted trees will benefit. By the time the ditch "melts" away after a couple of years, the vetiver hedge will be established and performing its function of increasing the infiltration of runoff, halting the loss of soil and soil nutrients, and creating a natural terrace.

Because the collection of runoff in the contour ditches has the effect of doubling or tripling the amount of effective rainfall, fruit trees planted by this method need no irrigation in the first three years of establishment. The vetiver grass lines stabilize the whole system. Photos 28 and 29 show the success of such systems.

Mulching
After the vetiver hedges are properly established, the farmer can cut down the vetiver grass to ground level when the dry season sets in and use its leaves as mulch at the base of the fruit trees to help retain stored moisture. (See Figure 30 and Photo 30). The advantage of using vetiver for this purpose is that its leaves harbor few insects and last well as a mulch. Vetiver hedges also protect the young plants in the hot summer months by providing some indirect shade; in the colder

winter months the hedges act as windbreaks.

Photo 28 Vetiver hedges protects tree crops, in this instance coffee in Costa Rica. The grass can be cut to mulch the coffee.

Photo 29 . Vetiver hedgerows protect citrus trees in Jiangxi Province, China - this area occasionally receives snow and is subject to low winter temperatures.

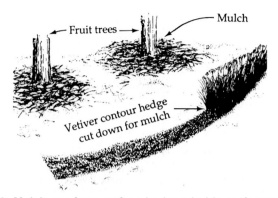

Figure 30. Mulch cut from vetiver hedges holds moisture for the fruit trees, repels insects, prevents fungal disease, and lasts longer than most other mulches.

Photo 30. On a coffee plantation in Ethiopia, vetiver mulch controls pests and improves moisture availability, soil organic matter and soil nutrients, all essential for high production.

Establishing forests

Forest trees should be planted by the same method as when establishing tree crops—on the contour and, if possible, in V ditches bordered by vetiver hedges to stabilize them. Where this has been done, the results have been spectacular: more than 90 percent of the seedlings planted by this method survived the 1987 drought in Andhra Pradesh, India, whereas 70 percent of the other seedlings died.

Stabilizing masonry walls for hill farming

In the Indian Himalayan highlands, where farming is carried out on terraces, vetiver grass is used to stabilize the masonry risers that have been erected over the centuries. The walls, pictured in the photo in Figure 31a, need all the support they can get to withstand the massive damage caused by runoff from associated torrential rain pouring over the walls, and even through the walls, from the catchment above.

Without some form of vegetative support, these ancient structures require continual maintenance. If one riser washes out during a heavy storm, other terraces farther down the slope often suffer considerable damage because of the domino effect. Photo 31 and Figure 31a, depict a typical terrace system in the hills, shows the type of damage frequently sustained. To allow for drainage between the stones, the masonry risers are not bound together with mortar. If the walls were solid, instead of just a small section falling out, the whole wall might collapse and trigger a landslide that could destroy the entire farm. Although these terraces have done an excellent job through the years, they do exact a toll in the form of crop losses, and they require a lot of hard work in repairs.

When we explained the Vetiver System of stabilization to the hill farmers, they wanted to plant as many areas as possible. In a World Bank project begun in 1986, vetiver grass was planted along the edge of the terraces during the rainy season in the hope that its strong root system would reinforce the masonry risers. It worked perfectly.

Figure 31c shows what the vetiver grass-protected terraces should look like once established. The grass is planted only at the extreme edge of each terrace so the hedges do not impede the essential drainage between the stones. According to the farmers, most of the damage during heavy storms is caused by the cascading of water down the slopes and over the top of the masonry terraces, especially if the water has a chance to concentrate into a stream. Once established, the vetiver hedges should take most of the erosive power out of this runoff as well as protect the edge of the terrace.

Photo 31. In the Himalayan highlands terraces are frequently washed out by concentrated flows of run-off water.

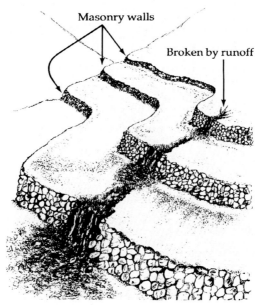

Fig 31a. A terrace system in the hills can easily be washed out during the rainy season.

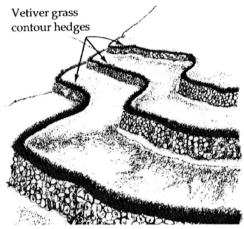

Figure 31b. Vetiver grass, planted on the extreme edge of each terrace, stabilizes the terraces without interfering with the essential drainage between the stones.

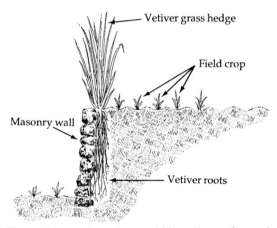

Figure 31c. The vetiver root system stabilizes the entire rock face of the vulnerable masonry risers.

As shown in the close-up in Figure 31c, the masonry risers are vulnerable because they are simply stones stacked on top of each other and are usually 2 to 3 meters high. Because its strong root system can easily penetrate to the bottom of the risers, vetiver grass can be used to protect the entire rock face.

In another project in the Himalayan highlands, in areas with no masonry terraces to halt massive sheet erosion, vetiver grass contour lines are being established to determine whether the natural terraces that build up behind the hedges will form a base of stable land for the production of fuel wood and fodder crops. In China, in the provinces of Jiangxi and Fujian, vetiver grass hedges are now being used to protect the edges of citrus and tea terraces.

Protecting roads.
Vetiver grass is also used to protect cut and fill batters, as shown in Figure 32. In the West Indies, the plant has been used extensively for stabilizing roadsides and has completely prevented erosion for years. People in St. Vincent in the Caribbean use it to line the outer edge of the tracks to their houses. The grass has exhibited a remarkable ability to grow in practically any soil. In Andhra Pradesh, India, for example, it was observed growing at the Medicinal and Aromatic Research Station at the top of a bare hill. Even though the soils on that hill are skeletal and granite boulders and had to be bulldozed to make a plot for the grass - and are deprived of most of the benefits from rainfall (since they are located at the very top of the hill), and at the time supported no other form of growth, the vetiver grass showed no signs of stress. A plant that can thrive under these extreme conditions should be able to do an excellent job of stabilization almost anywhere.

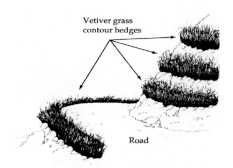

Figure 32. **Vetiver hedges stabilize roadsides.**

Photos 32a and 32b show how vetiver grass has been used in Malaysia to protect roads.

Photo 32a. In Malaysia, vetiver hedgerows stabilize low cost plantation roads at minimal cost and maintenance.

Photo 32b. In Malaysia, vetiver hedgerows stabilize what had been a very unstable highway fill on this high cost expressway.

In Photo 32a, the hedgerows stabilize the slopes of plantation roads. Once the hedges are established, the roads are fully stabilized and do not slip. The cost to establish the hedges is minimal, and the maintenance is even less. The same Vetiver System is used to stabilize high-cost highways, as shown in Photo 32b.

Stabilizing wasteland development
In the Sahel region of Africa (Niger and northern Nigeria) and in Bharatpur (central India), under the extreme conditions of constant fire and drought, *C. nigritana* and *C. zizanioides*, respectively, have survived as the climax vegetation for hundreds of years. The use of vetiver grass in wasteland development has proven effective as the initial stabilizing plant. When planted as contour hedges across wasteland areas - the first stage in wasteland stabilization - *C. zizanioides* reaps the benefits of surplus runoff and harvests organic matter as it filters the runoff water through its hedges. The improved micro-environment of improved soil moisture allows for the natural generation of native species between the vetiver hedgerows. As an example, the use of vetiver grass is in most instances an excellent alternative to an engineered structure as a measure to control complex multi-gullied land areas. The addition of a masonry plug at the end of the system allows silt to build up and gives the grass a basis of establishment. (See Figure 33).

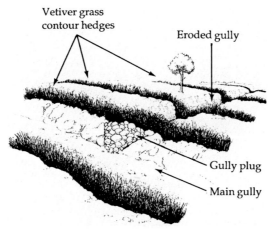

Figure 33. Vetiver hedges can stabilize multi-gullied wasteland areas.

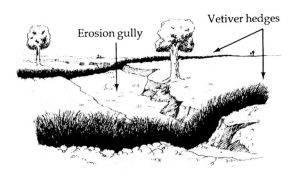

Figure 34. Vetiver hedges stabilize gullies.

The same would apply to smaller, up slope gullies as shown in Figure 34. Once established, the grass would fill and terrace the gullies with silt. In an upland area of Fiji, as shown in Photo 34, a gully was stabilized with vetiver grass some 30 years before the photo was taken. In the photo, the gully is fully stable and the hedges remain where they were originally planted.

Photo 34. In this upland area of Fiji some 30 years before this photo was taken, this gully was stabilized with vetiver hedgerows, that still remain where they were originally planted.

Stabilizing riverbanks and canal walls

Using vetiver grass to stabilize river banks and canal walls is another recommended practice. In an experiment in Tanzania, on the road to Dodoma, a road engineer used vetiver grass to protect the wing wall of a bridge on one side of the river and constructed the usual concrete wing wall on the other side. Some 30 to 40 years later, the concrete wall had already collapsed into the river, and the bank it was protecting was eroded. On the other side, the vetiver grass was still holding the bank in perfect shape. Figure 35 shows how vetiver grass can be used to protect the river approaches to a bridge. The success of the Vetiver System is evident in Photo 35, where newly planted vetiver hedges protect the approaches to a small bridge in El Salvador. During the floods of Hurricane Mitch in 1998, this bridge did not wash out.

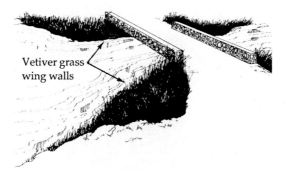

Figure 35. Vetiver hedges protect the river approaches to bridges.

Photo 35. In El Salvador, newly planted vetiver hedgerows protect the approaches to a small bridge.

Figure 36 shows how vetiver grass can be used to protect the bank of a major canal. In Vietnam, as seen in the photo 36, the canal has been stabilized with vetiver hedgerows, and the plants do not invade the river or the fields behind the rows.

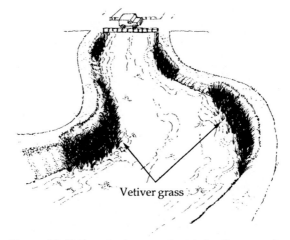

Figure 36. Vetiver hedges protect irrigation canals.

Photo 36. Vetiver hedgerows used to protect both sides of a canal bank in Vietnam. (Tran Tan Van).

The typical problem is depicted in the top illustration in Figure 36 the concrete conduit is undercut by erosion at point A and fills with silt at point B. To overcome this problem, vetiver grass should be planted parallel to the upper and lower sides of the concrete conduit. As shown in the bottom illustration, the upper hedge will prevent silt from entering the canal, while the lower two hedges will prevent erosion and thereby keep the concrete structure from being undermined by rills or gullies.

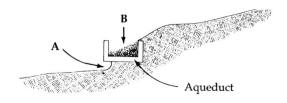

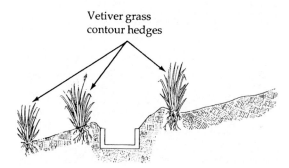

Figure 37. Vetiver hedges protect irrigation channels.

On a sugarcane farm in Zimbabwe, as seen in Photo 37, the left side top vetiver hedge is protecting an irrigation channel from silting up behind the hedge, while the other rows protect the drain from erosion and silting and help maintain its shape. In addition as the vetiver matures it acts as a "roof" above the drain and suppresses weed growth, and thus reduces maintenance costs.

Photo 37. On a sugar farm in Zimbabwe, a vetiver hedgerow protects an irrigation channel (located behind the far left hedge), while other rows protect the drain, and the hedge on the right protects the road.

Protecting dams

A similar approach can be taken to protect dams. Small dams are silting up at an alarming rate throughout the world. Once they become filled with silt, they are of no further use - and in many cases there is no other site suitable for a new dam. If vetiver grass is planted around the sides of the dam, as shown in the illustration below Fig. 38, the silt carried by runoff from the surrounding hills will be trapped before it reaches the dam. Vetiver hedges planted across the inlets (A) of small dams on intermittent streams will protect the dams from siltation. In time, these hedges will form stable terraces that can be used for cropping or tree planting. In the bottom illustration, vetiver has been planted on the walls of a dam to protect them from being worn down by rill erosion, a problem afflicting many unprotected earth dams around the world. A note worthy use of vetiver hedgerows is for stabilizing earth spillways of farm dams. This is a very effective way of preventing spillway erosion.

To make it easier to spot seepage along the toe, or very bottom of dam walls and canal banks, vetiver should not be planted in those areas. Photo 38a shows a small farm reservoir in Australia around which the farmer planted vetiver grass to protect the reservoir from wave action (lap erosion). Photo 38b shows a dam wall in Zimbabwe on which vetiver provides complete protection and stabilization.

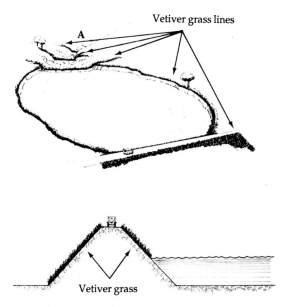

Figure 38. Vetiver hedges protect dams.

Photo 38a. Vetiver grass planted around the shore of a small farm reservoir protects the reservoir from wave action.

Photo 38b. Vetiver grass planted on this dam wall protects the down slope from erosion.

Common applications

The versatile vetiver plant has numerous common applications as well. It makes good bedding for livestock, because it soaks up the urine and stays dry longer. Ultimately, it makes good compost. In countries with strong winds, vetiver grass hedges make good windbreaks to protect vegetable gardens, young fruit and timber trees. Hedgerows when green also serves as a firebreak. Vetiver is used as mulch for tree crops and as pest proof thatch for roofs of houses, sheds, and shelters. (See Photo 39). The grass is woven into baskets, and other handicrafts, and the leaf midribs and flower stems make excellent brooms.

Photo 39. In Zimbabwe, vetiver grass thatch is popular. With good thatching techniques, vetiver thatch will last for many years.

Photo 40. Vetiver is a very good source of material for handicrafts. This hat was made by ladies in Colombia.

In more recent years we have found that vetiver can be used for a number of very useful non conservation purposes on a farm:

- Vetiver Systems has the ability to remove nitrates, phosphates, and BODs from farm animal effluent, bringing the pollutant levels within national environmental standards. This can be done by creating small constructed wetlands through which effluent is passed. Because vetiver grass can grow so fast under such conditions it will produce about 100 tons of dry matter per ha of high quality high protein fodder that can be used as fresh feed or hay.
- Vetiver when planted around ponds or grown on floating pontoons will help reduce nitrate levels in pond water and reduce the incidence of blue green algae, at the same time provide for forage.
- Vetiver grass will attract the maize stem borer moth - *Chilo partellus*. The latter prefers to lay its eggs on the vetiver leaves, the larvae when hatched fall off and die. Stem borer damage to maize is reduced significantly (as much as 90%). It may turn out that vetiver will have the same "push-pull" impact on rice and cane borers.

- As other non agricultural applications of the Vetiver System expand in other sectors, there is and will be a growing market demand for vetiver plant material. Farmers will then be able to establish nurseries and sell the plants by the slip. It is possible to produce 500,000 slips per ha per year, worth in India today (2008) US $12,500.
- Farmers who live near urban areas could also produce vetiver as an ornamental plant for sale in pots.

Thus there are many benefits to farmers from growing vetiver grass, and it is important to tell farmers about these alternative uses at the same time as encouraging them to use vetiver for soil and water conservation.

Management Tips

We asked users to give us their views and share their experiences. Below are some of the responses we received.

General observations

- Well-grown vetiver hedges result in less runoff and improved groundwater supplies. Dry-season stream flow improves under the hedge system of in situ moisture conservation.
- In most instances on slopes of up to 5 percent, about 10 centimeters of silt is deposited behind the hedges annually.
- In addition to its use for soil and moisture conservation, vetiver is being used for fodder, thatch, mulch, livestock bedding, windbreaks, roadside protection, pollution control, and brooms.
- Where hillside crop drainage is required—as in the case of tobacco ridges on a graded slope—vetiver hedges act as an excellent buffer against erosion if placed on the contour at fixed intervals on the hillside.
- Most vetiver plant roots grow straight down for at least 3 meters. Other plant roots will grow out into the field for up to 50 centimeters, but vetiver roots do not significantly affect crop growth—probably because of the high moisture content of the soil and a symbiont mycorrhiza associated with the hedge.

Vetiver hedges take about three years to be fully effective under low rainfall conditions. If vetiver slips are planted 10 to 15 centimeters apart, the hedge will form more quickly. Even where gaps exist, inter-plant erosion does not seem to be a problem because the roots join together in the first year to form a sub-surface barrier.

- Where vetiver is planted along the edge of terraces, forward-sloping terraces are better than backward-sloping terraces because less runoff is removed by the terrace back channels. Also because one can dispense with the back channel - and also in some instances the front channel, where constructed- more land will be available for cropping. The ultimate objective should be to dispense with man made terracing, where possible, through the use of vetiver hedges, so that the topsoil is not lost as part of runoff, but is filtered out to form a natural terrace.

- Vetiver has been observed growing under conditions ranging from 200 to 6,000 millimeters of rainfall annually and at 2,600 meters above sea level. It survives snow and frost (in tropical mountain areas) and grows on most types of soil. It obviously grows better where the soil is moist and fertile, but even under adverse conditions it grows extremely well compared with other grasses.

- In many countries vetiver has been infected with brown spot. The disease does not seem to have an adverse effect on its growth, however. A few instances of black rust have been observed but are not significant. In India the rust seems to be vetiver-specific and does not cross-infect other plants. In China stem borers have attacked vetiver, but in most cases the borer dies once it gets in the stem. Farmers generally are unconcerned and tend to respond by selecting plants that are more pest and disease resistant.

- Some early results from India, on both alfisols and vertisols, indicate that rainfall runoff was reduced from 40 percent to 15 percent (compared with the control), and silt loss was reduced from 25 tons per hectare to 6 tons per hectare (all for 2-year-old hedges on 2 percent slopes). The time to wilting in one demonstration on alfisols increased from 7 days to 20 when in

situ moisture conservation measures were applied.
- An interesting technique observed in China was the plaiting, or interlacing, of vetiver leaves and stems from separate, neighboring plants to create a temporary barrier until the full hedge could be established.
- The cost of vetiver hedges depends on the availability and cost of planting material. In India the initial cost of hedge establishment is estimated at US$8 per 100 meters of hedge, US$6 of which goes for planting materials and other inputs. Once the live material, in the form of a hedge, is on the farm, the cost to produce new hedges is relatively low—it may be as little as US$2 per 100 meters. Under such conditions the economic rate of return is more than 100 percent. Where the slopes are less than 5 percent and the hedges are spaced about 40 meters apart, 250 meters of hedge is required per hectare at a cost of between US$5 and US$20. (See Table 7-3 at the end of this chapter.)

Selecting planting material

- In Karnataka, India, to date six cultivars have been identified. One cultivar selected over the years by farmers exhibits superior characteristics for hedge formation; fodder; and insect, disease, and drought resistance.
- When selecting material, choose vigorous plants that exhibit resistance to pests and diseases and that tiller well.
- Where winters are cold, select material that is more tolerant of cold temperatures.

Establishing nurseries

- Vetiver planted densely across large gullies can be used for replanting elsewhere. Gullies make good informal nurseries because often they are permanently moist and have conditions good for growth.
- Stem and root cuttings grown under plastic may be a cheap way of vegetative propagation.

- For optimum tillering, nurseries should be fertilized (150 kilograms per hectare of nitrogen) and irrigated (especially in very dry areas).
- Nursery plants should be cut back to about 30 to 50 centimeters to encourage tillering.
- The best nurseries seem to be in loamy sands to sandy-clay soils where the drainage is good and where it is easy to dig up the plants for transplanting. We have seen excellent nurseries (when well watered) in sandy areas near perennial rivers.

Field Planting

- As long as the vetiver is planted when the ground is wet, it can survive a long period of drought after planting.
- On very small farms and fields where land is scarce and where farmers are reluctant to plant across their fields, vetiver should be planted on the field boundaries.
- On non-arable lands that are heavily eroded, vetiver should be planted first across the gullies and around the gully heads. The excess material from the gullies can then be used for planting across the slopes in subsequent years.
- Gap filling is essential and should be done at the beginning of the wet season. The possibility of "layering" live stems across the gaps should be tried as a gap-filling measure.
- To encourage tillering and hedge thickening, the grass should be cut back to 30 to 50 centimeters after the first year. Cutting in the first year does not seem to have any incremental impact on tillering.
- Termite infestation (attacking dead material) can be controlled by applying 1 kilogram of benzene hexachloride (BHC) for every 150 meters of hedge line.
- Once the vetiver has established (one month after planting), plowing a small furrow immediately behind the vetiver hedge line helps to capture runoff and results in better growth of the plant.

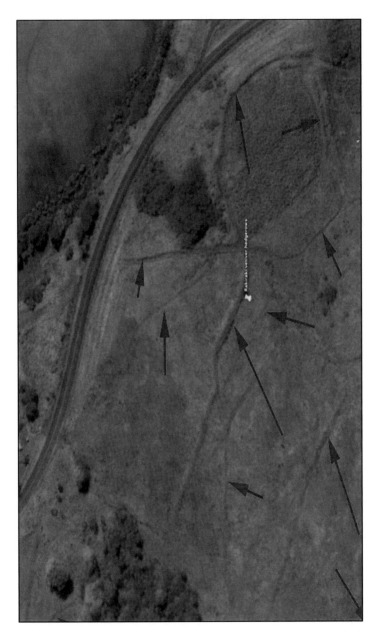

Photo 41. A 2007 Google Earth image of farmland in Fiji that John Greenfield planted with vetiver hedgerows in the 1950s'. The hedgerows are still there (red arrows) after 50 years.

Table 1. Slope, Surface Run, and Vertical Interval

Slope		Gradient			Surface run[a] (meters)
Degrees	Percent				
1	1.7	1	in	57.3	57.3
2	3.5	1	in	28.6	28.7
3	5.3	1	in	19.1	19.1
4	7.0	1	in	14.3	14.3
5	8.8	1	in	11.4	11.5
6	10.5	1	in	9.5	9.6
7	12.3	1	in	8.1	8.2
8	14.0	1	in	7.1	7.2
9	16.0	1	in	6.3	6.4
10	17.6	1	in	5.7	5.8
11	19.4	1	in	5.1	5.2
12	21.3	1	in	4.7	4.8
13	23.1	1	in	4.3	4.5
14	25.0	1	in	4.0	4.1
15	27.0	1	in	3.7	4.0
16	28.7	1	in	3.5	3.6
17	30.6	1	in	3.3	3.4
18	32.5	1	in	3.1	3.2
19	34.4	1	in	3.0	3.1
20	36.4	1	in	2.8	3.0
21	38.4	1	in	2.6	2.8
22	40.4	1	in	2.5	2.7
23	42.5	1	in	2.4	2.6
24	44.5	1	in	2.3	2.5
25	46.6	1	in	2.1	2.4
26	48.8	1	in	2.0	2.3
27	51.0	1	in	2.0	2.2
28	53.2	1	in	1.9	2.1
29	55.4	1	in	1.8	2.1
30	57.7	1	in	1.7	2.0
31	60.1	1	in	1.7	2.0
32	62.5	1	in	1.6	1.9
33	65.0	1	in	1.5	1.8
34	67.5	1	in	1.5	1.8
35	70.0	1	in	1.4	1.7
36	72.7	1	in	1.4	1.7
37	75.4	1	in	1.3	1.7
38	78.1	1	in	1.3	1.6
39	80.1	1	in	1.2	1.6
40	84.0	1	in	1.2	1.6
41	87.0	1	in	1.2	1.5
42	90.0	1	in	1.1	1.5
43	93.3	1	in	1.1	1.5
44	96.6	1	in	1.0	1.4
45	100.0	1	in	1.0	1.4

a. The figures for the surface run are based on a vertical interval (VI) of 1 meter. To use this table, multiply the surface run by the VI; for example, with a VI of 2 meters on a 70 percent slope, the surface distance between vegetative barriers = 2 x 1.7 = 3.4 meters.

Table 2. Cost of Land Treatment with Contour Hedges of Vetiver Grass by Slope Classification and Cost of Labor

(U.S. dollars per hectare)

Slope (percent)	Daily cost of labor					
	$0.50	$1.00	$1.50	$2.00	$2.50	$3.00
0–1	2.43	3.44	4.45	5.46	6.47	7.48
1–2	7.29	10.32	13.35	16.38	19.40	22.43
2–5	17.02	24.08	31.15	38.21	45.28	52.34
5–10	36.46	51.60	66.74	81.88	97.02	112.17
10–15	60.77	86.00	111.24	136.47	161.71	186.94
15–20	85.08	120.40	155.73	191.06	226.39	261.72
20–30	121.54	172.01	222.48	272.95	323.42	373.89
30–40	170.15	240.81	311.47	382.12	452.78	523.44
40–50	218.77	309.61	400.46	491.30	582.15	672.99
50–60	267.38	378.41	489.45	600.48	711.51	822.55
60–70	316.00	447.22	578.44	709.66	840.88	972.10
70–80	364.61	516.02	667.43	818.84	970.25	1,121.66
80–90	413.22	584.82	756.42	928.02	1,099.61	1,271.21
90–100	461.84	653.62	845.41	1,037.19	1,228.98	1,420.77

Note: Figures are based on a vertical interval of 2 meters.

Made in United States
North Haven, CT
12 February 2022